BROCK UNIVERSITY LIBRARY

Presented by

Mr. William K. Bailey
Toronto
January 1991

THE RARE
BOOKS OF FREEMASONRY

THE BOOKMAN'S JOURNAL
AND PRINT COLLECTOR
Monthly 2/-

AN INTERNATIONAL ORGAN FOR COLLECTORS OF BOOKS & PRINTS

"The Bookman's Journal and Print Collector" Company, Ltd.

AT THE SIGN OF THE OPEN BOOK

London: 173-5 Fleet Street, E.C.4

New York: R.R. Bowker Company

THE RARE BOOKS OF FREEMASONRY

By
LIONEL VIBERT
P.M. Lodge Quatuor Coronati,
Author of *The Story of the Craft*, etc.,
Editor *Miscellanea Latomorum*.

London:
"The Bookman's Journal" Office
173-5 Fleet Street, E.C.4
1923

PREFACE

A carefully compiled list of the rarer masonic books has been required for a long time past, and few students of the literature of the Craft have had either a competent knowledge of the scarce works, or leisure to enable them to investigate and describe in a methodical manner the rarities which in the present work are brought under review. Fortunately the enterprise of the publishers has secured the services of Mr. Lionel Vibert, who as Editor of *Miscellanea Latomorum* and author of some charming little volumes on old-time masonry, brings to bear on his present task both knowledge of masonic literature and care in tracing out and marshalling for our benefit the facts here embodied in a concise and orderly form.

The masonic student will find this work of material value for reference and comparison; the collector will likewise have a competent authority to consult; while the beginner will probably be encouraged to investigate more closely that class of scarce books which might otherwise never have been brought to his notice. To these and all others interested, the success of Mr. Vibert's former works is a guarantee that they have now available a descriptive catalogue of the rare books of Freemasonry which can be relied upon for its accuracy and comprehensiveness.

WM. WONNACOTT
Librarian of the Grand Lodge

July 1923.

CONTENTS

	Page
Preface	5
Introduction	9
(a) Constitutions	11
(b) Pocket Companions	20
(c) Exposures	24
(d) Historical	29
(e) Sermons and Speeches	33
(f) Miscellaneous	37

INTRODUCTION

THE Grand Lodge of England was formed in 1717, but did not attract any degree of public attention till 1721; up to which time the Freemasons had been so inconspicuous that there are not known to exist more than fifteen distinct references to them in print of an earlier date. These are collected in the appendix to the Inaugural Address of Mr. E. H. Dring, which he delivered to the Quatuor Coronati Lodge in 1912 (to which the present article is very much indebted), and of them only three can be said to be anything more than passing references to the Craft and its customs.

It was not till 1722 that any work was printed which can fairly be described as a work on Freemasonry. This was the pamphlet known to-day as the *Roberts Constitutions*. The Freemasons had possessed in manuscript, ever since the fifteenth century at all events, documents containing what purported to be a history of the Order, and a set of ordinances comparable to those of a Gild. Versions actually made in the fifteenth, sixteenth and seventeenth centuries have come down to us to-day, and of these there are nearly a hundred in existence. They are known as the Old Charges, and the versions are almost all so closely allied textually as to make it obvious that they derive from a single original. They begin with a prayer, the opening words of which are usually: "The Almighty Father of Heaven"; and then, after an introduction beginning "Good Brethren and Fellows, our purpose is," there follows the historical section, and the Charges themselves in two or more sections. Even to-day versions are still occasionally discovered, and the great majority of those extant were written in the seventeenth and eighteenth centuries.

The *Roberts Constitutions* was a printed version of such a manuscript, and of this pamphlet until quite recently only one copy was known to exist, which is in the library of the Grand Lodge of Iowa. A reprint was issued by Messrs. Spencer's, of Great Queen Street, in 1871.

The title-page of the original is:
The / Old Constitutions / Belonging to the / Ancient *and* Honourable / SOCIETY / of / Free *and* Accepted / MASONS. / *Taken from a Manuscript wrote about Five / Hundred Years since.* / LONDON : / Printed and Sold by J. Roberts, in / *Warwick Lane*, MDCCXXII. / (*Price Six-Pence.*)

and the collation is: Title, reverse blank; four pages not numbered, preface; pp. 1-26 History and Charges, but the last two pages are erroneously numbered 23, 24.

The next half-century saw the Craft established on the Continent, and two rival Grand Lodges working in London itself. But the output of books was not profuse, and such as there were consisted largely of controversial pamphlets, alleged exposures and reprints of speeches and sermons. Still, many of the items of early date are extremely rare, but, as it was not till somewhere in the eighties of last century that there was any market for early masonic literature, the prices of an earlier date, when we can ascertain them, are no criterion of value. To-day there is a far larger public interested in the subject and prices are rising.

Wolfstieg in his monumental Bibliography, published in 1912 and containing well over 43,000 entries, recognises eight classes with 174 sub-divisions. It will be sufficient here to deal with the subject under the six heads: Constitutions; Pocket Companions; Exposures; Historical; Sermons and Speeches; and Miscellaneous.

(a) CONSTITUTIONS

THIS is the title given by Grand Lodge since 1723 to the official publication which to-day comprises certain "Ancient Charges" and the Laws of the Craft. The work was first issued in 1723, and up to 1897 twenty-four editions are recognised. The idea was copied, and the actual work also reprinted in Ireland, Scotland and America.

The name *Constitutions* was also that given in many cases to the manuscripts of an earlier date, and as several of these were now printed these works will also come as a sub-division under this heading. Taking this class first we have:

(Part I)

(1)

1722. The Roberts Constitutions of 1722 already described. To the copy belonging to the G.L. of Iowa, hitherto considered unique, has now to be added a copy which recently passed through the hands of Messrs. Fletcher, of Bayswater, and is now in private ownership.

(2)

[1724]. The Briscoe Print. It does not seem necessary to give the elaborate title in full; the first part of it is:
> The secret / HISTORY / of the / Free-Masons / being an / Accidental Discovery / of the / Ceremonies Made Use of in the several / LODGES, /

and it also contains Observations . . . on the New *Constitution* Book . . . written by James Anderson (*vide infra*, No. II., 1.); and "a short dictionary of private Signs, or Signals." Collation: Title with reverse blank; iv pp. preface; 47 pages of text with reverse of p. 47 blank. The Imprint is: London; Printed for Sam. Briscoe, at the Bell-Sauvage, on Ludgate Hill (etc.); but with no date. Mr. G. W. Bain reproduced his copy in facsimile in 1891; there may be other copies in private ownership but no library appears to possess one.

(3)

1725. The Briscoe Print, second edition. Identical except for the date on title-page. Also of great rarity; there is a copy in the Worcestershire Masonic Library.

(4)

[1729]. Cole's Constitutions. This consists of a Dedication to Lord Kingston, the title as follows:

A / Book / of the Antient / Constitutions / of / the Free & Accepted / MASONS both being engraved, and fifty-one engraved copper plates of text, followed by six more of songs. Kingston was Grand Master from December 27, 1728, to December 27, 1729, so that the work may be presumed to have been issued in 1729. It is found with 38 added pages in ordinary type (two speeches, a Prologue and an Epilogue), and may have been re-issued with this addition in 1730, when the combined work was advertised (on March 18) as just published. There is in the Library of Grand Lodge a specially prepared copy on a paper of a larger size.

(5)

1731. Cole's Constitutions, second edition. This consists of a Frontispiece, a very lengthy title-page with imprint and date, and the original plates; but with the Dedication altered so as to read to Lord Lovel, and the imprint obliterated from the title. The six copper plates of songs are also omitted. Then follow the speeches and Prologue and Epilogue, pp. 1—34 and two not numbered, and two collections of songs, together occupying 64 pages. But the order of binding differs in different copies. The speeches have their own pagination and title-page, which describes them as the second edition with date 1734; while the songs are also separately paginated with a title-page dated 1731. Wolfstieg gives the first edition a lengthy title-page almost identical with that of 1731, and with the date 1728. This does not correspond with the descriptions given by Hughan in his introduction to the Jackson reprint of the second edition (Leeds, 1897) and by Mr. Dring (*op. cit.*); it may possibly refer to the issue of 1730.

(6)

1751. Cole's Constitutions, third edition; a reprint of the former but all in ordinary type. This has a new and much shorter title-page. 8vo; 78 pages; and two plates.

(7)

1762. The fourth edition; similar.

One masonic library in this country (Worcestershire) possesses the quartette. But it may be mentioned, as illustrative of how the prices actually obtained for these things are no sort of guide to their intrinsic value or their probable selling price to-day, that the set of four was in the Spencer Sale of 1875 and went for £3 8s.

(8)
1739. The Dodd Version. Title :—
> The / BEGINNING / and / First Foundation / Of the Most Worthy / Craft of Masonry / with / The Charges thereunto belonging. / By a Deceas'd Brother, for the Benefit of his Widow / LONDON : / Printed for Mrs. Dodd, at the Peacock without Temple Bar / mdcc xxxix. (Price Sixpence)

4to ; 20 pages. This is the last of the printed versions of the old manuscripts, and is even rarer than the Cole. Hughan (*Old Charges*, 1895, p. 139) says : " I think there must be four at least in existence." Nevertheless in the Spencer Sale the copy then offered as one of the only three known fetched 23/-. The text is all but identical with that of the Cole.

We can now consider the series of *Constitutions* which were issued by, or with the approval of, Grand Lodge, and the reprints of them. We have :—

(Part II)

(1)

The Constitutions of 1723. The full title is :
> The / CONSTITUTIONS / of the / FREE-MASONS. / containing the / History, Charges, Regulations, &c. / of that most Ancient and Right / Worshipful FRATERNITY. / For the Use of the Lodges. / London : / Printed by William Hunter, for John Senex, at the *Globe*, / and John Hooke at the *Flower-de-luce* over-against *St. Dunstan's / Church*, in *Fleet-street*. / In the Year of Masonry —— 5723 / Anno Domini —— 1723 /

Collation : half-title—Constitutions between ornamental borders ; plate ; title as above, reverse blank ; dedication in large type on two leaves not numbered ; text pp. 1—74; songs with music pp. 75—90 ; p. 91 contains a notice about the rest of the music, the license to publish of the Grand Master, and FINIS. The reverse of p. 91, not numbered, has publishers' announcements. Perfect copies with the half-title are very rare.

(2)

1725. The *Constitutions* of 1723 re-printed in Dublin. No copy is known to exist ; the work is only known from an advertisement in *The Dublin Journal* of July 31, 1725.

(3)

1730. Pennell's *Constitutions*. Anderson's title is repeated, but the imprint is :—
   DUBLIN / Printed by *J. Watts*, at the Lord *Caterets* / Head in *Dames-Street*, for *J. Pennell*, at the / three *Blue Bonnets* in *St. Patrick's-Street*. / *In the Year of Masonry* 5730 / *Anno Domini* 1730.
   4to; plate and 96 pages. The work is Anderson more or less reproduced, but with some added matter. Dr. Chetwode Crawley only knew of one perfect copy, in private ownership in America.

(4)

1734. Franklin's reprint of Anderson in Philadelphia " by special order, for the use of the brethren in North-America 1734."

(5)

1738. Anderson's second edition. Described as the New Book of Constitutions, by James Anderson, D.D., in a long title-page. Collation: plate; title-page; reverse blank; pp. i—x, dedication and preface; two pages not numbered, sanction to publish, plate with the arms of Carnarvon and his titles; pp. 1—230 text; two pages not numbered, corrigenda and publisher's announcements. The songs conclude on p. 215, and are followed by a reprint of a pamphlet (No. F. 7 *infra*) " A Defence of Masonry, published A.D. 1730. Occasion'd by a Pamphlet call'd *Masonry Dissected* " (as to which *vide infra* No. C. 3). This also appears in a *Pocket Companion* of this same year (*vide infra* No. B. 4). This is in turn followed by *Brother Euclid's letter to the Author against Unjust Cavils*, which is dated 1738, the authorship of which is unknown. The leaf pp. 129—130 is substituted for an original which contained various errors, the most conspicuous of which was the writing STEPHEN instead of FRANCIS, Duke of Lorraine. No copy appears to be known with the original leaf *in situ*, but in one copy, at present in private ownership, it is found attached to the cover. The Plate is the same as that of 1723, with the exception of the lettering at bottom " Engraved by John Pine in Aldersgate Street, London," which is now deleted. This Plate measures $8\frac{7}{8}'' \times 7\frac{3}{8}''$. The work was printed—with identical typing—on paper of two different sizes, $8\frac{7}{8}'' \times 7\frac{1}{4}''$ and $7\frac{1}{4}'' \times 5\frac{1}{2}''$. The small paper copies could not therefore have taken the Plate without folding, and as none appear to be known that have it, it would in fact seem that none were in fact issued with it. Mr. Hughan, in his Preface to the facsimile issued by Lodge Quatuor Coronati in 1890, stated that he had only succeeded up to that time in tracing twenty-six copies of this edition. The imprint is :—

    LONDON : / Printed for Brothers Cæsar Ward and Richard Chandler, / Booksellers, at the *Ship* without *Temple-Bar ;* and sold at their / Shops in *Coney-Street*, YORK, and at SCARBOROUGH-SPAW. / MDCCXXXVIII. / In the *Vulgar* Year of Masonry, 5738.

(6)
1746. The / History and Constitutions / of the / Most ancient and honourable Fraternity / of / Free and Accepted MASONS : / (etc.). This is a reissue of the previous work, obviously the publisher's remainders, as it is identical in all respects, save that it has a new title-page beginning as above and with the imprint :—

    LONDON : Printed ; and sold by J. Robinson, at / the *Golden-Lion* in *Ludgate-street*. / In the vulgar Year of MASONRY 5746.

As before some copies were small paper, and were apparently originally issued without the Plate. It is still rarer than the 1738 edition ; Mr. Hughan in 1890 only knew of nineteen copies.

(7)
1751. A second edition of Pennell, (No. 3 *supra*), with a long title which concludes :

    Collated from the Book of Constitutions published in England, in the year 1738, by our worthy Bro. James Anderson. For the Use of the Lodges in Ireland. By Edward Spratt. Dublin. . 1751.

4to ; plate ; viii, 172 pp. A very rare work, almost as much so as the first edition.

The third edition of the Book of Constitutions was published in 1756 with a frontispiece by B. Cole. The author was Entick. He used the enlarged history that Anderson had written for his 1738 edition, but the Regulations were entirely recast. The fourth edition was published in 1767. It is described as " A New Edition, with Alterations and Additions, by a Committee appointed by the Grand Lodge." Entick's name still appears (as well as Anderson's), but in fact he had nothing to do with it.

(8)
1776. The Appendix to the fourth edition. This, written by William Preston, is rarely met with. It occasionally occurs bound up with the fourth edition. It consists of lxxvi pages. Page i has a half-title: Appendix / to the / Constitutions / of the / Society of Free and Accepted Masons. Page ii has the Resolution ordering the Appendix. Page iii commences " Appendix " and the text goes to lxx; then come an anthem and ode, and lxxvi, not paged, is blank.

Of the fourth edition there were two unauthorised issues, one in 8vo, published by G. Kearsly, London, and the other, also 8vo, but with another title-page and plates, published by Thomas Wilkinson, Dublin. Both appeared in 1769.

The fifth edition, of 1784, is described as "A New Edition, revised, enlarged and brought down to the year 1784, under the direction of the Hall Committee, by JOHN NOORTHOUCK." The frontispiece shows the interior of Freemasons' Hall, and is dated 1786. The usual collation is: Frontispiece; page not numbered "Explanation of the Frontispiece," reverse blank; i—xii, dedication, laws relating to the Charity, sanction, preface, contents; pp. 1—76, 77—134, 135—204, 205—350, 351—414, history in five parts (including Charges, Regulations, etc.); 415—459 poetry. On p. 459 is a note to the binder regarding the cancelled leaf. This was pp. 67, 68; the original p. 67 is headed "In Italy," the corrected page being headed "Gothic Architecture." The cancelled leaf is rare, and although the work was issued originally presumably without the plate, there do not seem to be any copies bound up without it, although some omit the laws relating to the Charity (*).

In 1753 a rival Grand Lodge had been founded in London which also issued an official publication with the title Ahiman Rezon. Of the eight editions issued the first four are rare; indeed, of the fourth itself only two copies appear to be known.

(9)
1756. Ahiman Rezon. 4to. A long title of which the commencement is:—

AHIMAN REZON: / or, / A Help to a Brother;

---

* A copy with the plate was sold at Sotheby's in 1908, for £1 3s. 0d.; but whether it contained the original cancelled leaf or not does not appear.

/ Shewing the / EXCELLENCY of SECRECY, / And the first Cause, or Motive, of the Institution of / FREEMASONRY ; /
and the imprint :—
LONDON : / Printed for the Editor, and sold by Brother *James Bedford*, at the / *Crown* in St. *Paul's Church-Yard*. / MDCCLVI.
The work was written by Laurence Dermott, the Secretary of this Grand Lodge, and the preface contains sarcastic references to the current histories of the Craft with allusions to other works, not all of which can now be traced. Collation: title; i—iii dedication; iv blank; v—xvii The Editor to the Reader; blank page not numbered; four pages not numbered, subscribers' names; four not numbered, contents; 1—208 and one page not numbered, text, headed Ahiman Rezon; songs preceded by a title-page with blank reverse; prologues and epilogues, and an Oratorio.

(10)

1764. Ahiman Rezon, second edition. An ornate title-page in a border with much less text and the imprint :
Printed for the Author / *and sold by* Br. ROBERT BLACK. / Bookbinder & Stationer / *in George Yard, Tower Hill*. / LONDON, 1764.
The text differs considerably from that of the first edition. There is a plate of the Arms of the Masons and those of the Operative or Stone Masons.

(11)

1778. Ahiman Rezon, third edition. The title-page is:
AHIMAN REZON : / or a / *Help* to *all that are, or would be* / Free and Accepted Masons. / (With many ADDITIONS) / The THIRD EDITION. / By Lau. Dermott, D.G.M. / (four lines of verse) / Printed for / JAMES JONES, Grand Secretary, / and Sold by / PETER SHATWELL, in the Strand, / LONDON, 1778.

(12)

1787. Ahiman Rezon, fourth edition. Similar but the publisher is Frakins.
The remaining editions were issued in 1800 (v); 1801 (vi); 1807 (vii); and 1813 (viii); the last two have lists of Lodges.
After the Union of the two Grand Lodges steps were taken to issue an official Book of Constitutions for the

united body. The Regulations were drafted by a joint committee, and as to publishing these there was no particular difficulty. But the history was another matter as the two bodies had taken very divergent views on that subject. Accordingly the historical section, or " first part," was held over, the Charges and Regulations being issued as a " second part," and this description " second part " was maintained for three editions but was then dropped, and no official history has ever been promulgated. But inquiries are still occasionally made for the non-existent first parts of the sixth, seventh and eighth editions. The first edition after the Union was that of 1815.

(13)

1815. Constitutions, sixth edition. Title-page:
CONSTITUTIONS / of the / Antient Fraternity / of / Free and Accepted Masons. / Part the Second / containing / The Charges, Regulations, / &ct., &ct. / Published, by / the Authority of the United Grand Lodge, / by / William Williams, Esq. / *Provincial Grand Master for the County of Dorset.* / LONDON : / Printed by W. P. Norris, Printer to the Society, / Little Moorgate, London-Wall. / MDCCCXV.

On the reverse of the title-page was a notice that the First Part would be printed with as little delay as possible, and that the laws would be revised after three years when sheets in which alterations had been made would be reprinted and sent to subscribers. The edition in its original state is rare.

The seventh edition was issued in 1819, with the title-page, foreword, and sanction of the previous edition unchanged, but there was now added a Preface to the Corrected Edition, dated from Belmont House, 19 February, 1819. Apart from the variations in the actual text this Preface alone serves to distinguish this edition. This was the last edition to be issued in 4to.

In 1823 the Provincial Grand Lodge of Upper Canada reissued this edition, with the same title-page except for the imprint, which now read :
First Canadian Edition. / Republished by order of the Provincial Grand Lodge of / Upper Canada. / Kingston: / Printed by H. C. Thomson. / MDCCCXXIII.

There was also a dedication which occupies the second leaf. Of this reissue there is a copy in the Library of the Supreme Council at Washington, D.C.

Of the later editions, the eighth, of 1827, was the last to retain the description "Second Part"; but only three of those of more recent years can be styled rare. These are :—

(14)
1855. 12th edition, the 32mo issue. (Also issued in 8vo.)

(15)
1865. 16th edition. 32mo. This is so rare that its very existence was doubted until quite recently, and it is only within the last year that Grand Lodge has become possessed of a copy.

(16)
1866. 17th edition, another 32mo, issue.

* * *

The Bain Reprint of the *Briscoe Print* has already been referred to. Of the *Constitutions* of 1723 complete or partial facsimiles have been issued at New York in 1855 and 1905, at Philadelphia in 1906, and at Wiesbaden in 1900. Of the 1738 edition an absolute facsimile was issued in 1890 as vol. VII of the *Quatuor Coronati Antigrapha*, with an introduction by Mr. Hughan. Of *Ahiman Rezon* there are as yet no facsimiles. But there is one work of this class which has to be reckoned among our Rare Books to-day, and that is :—

(17)
1869. Constitutions of the Freemasons. By William James Hughan. London : Spencer and Co. 8vo. pp. xxii, 38, 51. The Introduction deals with the Constitutions generally, and the Text consists of a reproduction of those of 1723, with the exception of the historical portion and the songs, and of the Cole Text. Only 70 copies were printed.

(b) POCKET COMPANIONS

ANDERSON'S first edition was exhausted by the beginning of 1735, and to meet what was no doubt a fairly extensive demand there was now published a work styled *A Pocket Companion for Freemasons*, the greater part of which was in fact simply a piracy of Anderson. This was the first of a series of similar works, which mostly contained a history, the Charges and Regulations, and other matter of the same kind, collections of songs and poetry, and lists of Lodges. After No. 1 in the list I now give, they usually have very long title-pages. Wolfstieg gives most of these fairly fully; I give in full one that he does not reproduce. They are all of considerable rarity in anything like perfect condition.

(1)

1735. Smith's Pocket Companion. Title :—
A / Pocket Companion / for / Free-Masons. / Deus Nobis Sol & Scutum/ Dedicated to the Society. / London : / Printed and sold by E. Rider in Blackmore- / street, near Clare-Market. / mdccxxxv.
The collection of songs, etc., is preceded by a sub-title with the date 1734 (p. 47).

(2)

1735. The same work printed at Dublin. Probably this was done with Smith's permission. This edition has a long title-page and a plate; and whereas the London edition was condemned by G. L. at London, where Anderson was still there to protest and was preparing his own second edition, the Dublin reprint was sanctioned and adopted by the Grand Lodge of Ireland.

(3)

1736. Smith, a re-issue. A long title-page and plate; printed by Torbuck.

(4)

1738. Smith, the second edition, " with large additions." Again, a long title-page, but no plate.

(5)

1752. A Pocket Companion, published at Edinburgh, of which Wolfstieg gives no details further than that a copy is in private ownership in Germany.

(6)

1754. (Entick). Scott's Pocket Companion, published by Baldwin; 8vo, viii, 328; with plate. The author's name does not appear, but he takes the course that he was again to take in the official third edition of the Constitutions (*supra*) of discarding the confusions of Anderson's 1738 edition.

(7)
1759. Scott, a second edition.

(8)
1764. Scott, a third edition.

(9)
1761. The Edinburgh Pocket Companion. Published by Auld, 8vo, with a long title.

(10)
1763. A second edition of No. 9; I give the title-page in full as a specimen of their general style :—
The / Free-Masons / Pocket-Companion / containing /The History of Masonry from the Creation to / the present Time ; / The Institution of the Grand Lodge of Scotland ; / Lists of the Officers of the Grand Lodges in / England & Scotland / WITH / A Collection of Charges, Constitutions, Or / ders, Regulations, Songs, &ct / The Second Edition / Edinburgh. / Printed for Alexander Donaldson / and sold at his shops in London and Edinburgh / mdcclxiii 8vo. ; vi, 274, pp.

(11)
1765. A third edition. Published by Auld and Smellie.

(12)
1764. A Pocket Companion. Published at Belfast by Magee, and described as the fifth edition.

(13)
1765. Another. Published at Glasgow by Galbraith.
There is another Glasgow edition of 1771, and others elsewhere of later dates, but none that can strictly be called rare. But there is one other rare work that comes into this category, although it is not styled a Pocket-Book and its contents are of a somewhat different character. It is :

(14)
1775. The Free-Masons-Calendar, or An Almanac for the year of Christ 1775, and Anno Lucis 5775. Containing, besides an accurate and useful calendar of all remarkable occurrences for the year, many useful and curious

particulars relating to Masonry. Inscribed with great respect to Lord Petre. By a Society of Brethren. London : Company of Stationers 1775.

* * *

It will be observed that these works nearly always contain a List of Lodges, and this will be a convenient place in which to notice the publications known as the Engraved Lists. An official List of Lodges was drawn up by Grand Lodge in 1723, but it was also necessary for each individual Lodge to possess at all times an authorised List, corrected to date, and accordingly in this same year the first Engraved List was issued, and they were continued up to 1778, one or more appearing annually. Until 1741 the engraver was John Pine who had done the Frontispiece to the *Constitutions* of 1723. In 1744 the engraver was Eman. Bowen, and from now onwards the engraver was either William or Benjamin Cole. These Lists are all more or less in the same style ; they are engraved on plates measuring six inches or slightly less by $2\frac{1}{8}''$, and each plate contains spaces for 12 Lodges, but blanks occur where Lodges on previous Lists have ceased to exist. The sign of the tavern where the Lodge meets is usually depicted, or the name, or both, and the street if in London, as also the days of meeting and the date of constitution. In the Lists from 1729 onwards a serial number is assigned to each Lodge and blank. The Lists for the earlier years are very scarce. Indeed as yet none are known to have survived for 1726-8, 1730-3, 1742, 1743, 1746, 1748 or 1749; and all other issues, except three, previous to 1765 are at present represented by single copies, and the same is the case with regard to several later ones.

The Grand Lodge of 1753 issued similar Lists but only two are known to exist, both for the year 1753, the earlier being in the Library of Lodge Quatuor Coronati, and the later at the South Kensington Museum. These are engraved on plates measuring $4\frac{5}{8}'' \times 2\frac{3}{8}''$, which have two Lodges to a plate, the information as to each being enclosed in ornate borders of Chippendale design, almost all different. There is also a title-page on a separate plate with an elaborate design of its own. The earlier List is reproduced in full in vol. xix of the *Transactions* of Lodge Quatuor Coronati ; the engraver was Jeremiah Evans, who does not however seem to have been a person of any note in his profession.

In 1775 Grand Lodge commenced the issue of an annual Calendar and List of Lodges and this superseded the Engraved Lists, the series having been continued till the present day. But no copy of the Calendar for 1816 appears to be in existence; at all events none was known to Mr. Lane when he published his *Masonic Records*, which puts in tabular form the information as to every known Lodge in the English Craft.

(c) EXPOSURES

IT being understood that the Society which had so suddenly sprung into prominence possessed secrets and practised ceremonies of an esoteric character, it was not long before persons of enterprise, if not of honesty, began to print alleged revelations of these mysteries. That they were at once denounced by the Fraternity as false and inaccurate would perhaps not by itself discredit them for us to-day; but, in fact, their interest now is purely antiquarian. They offer a possible basis for investigations as to what was the practice of the period in these matters; but if it at all resembled the accounts given in the various exposures it must have been greatly metamorphosed in the intervening years. Some were mere broadsides, but for the sake of completeness it is desirable that they also should be described. Accordingly, taking this class first, we have:—

(Part I)

(1)

1725. The Whole Institutions of Free-Masons Opened, (etc.). Printed by William Wilmot on the Blind-Key, 1725. This consists of a folio sheet printed on both sides. No place of publication is mentioned. Only one copy is known, which is in private ownership.

(2)

1726. The Grand Mystery laid open, or the Free Masons Signs and Words discovered. Printed in the year 1726. This is another single folio sheet, but printed on one side only. Again, only one copy is known, which is in the possession of the same owner as No. 1.

(3)

1730. The Mystery of Freemasonry.
This was a broadside which was reported as being "lately published and dispersed about the Town," and which was reprinted in the *Daily Journal* of Aug. 15, 1730. It was again reprinted in the issue of Aug. 18. No specimen of the original appears to be known, but single copies of an engraved broadside with the same text, which is apparently a separate and later publication, are in the B.M., and three masonic libraries. It was also reprinted with the heading:—

The Mystery and Motions of Free-Masonry discovered. London, Printed by Edward Nash, in King Street, Covent Garden. mdccxxx.

Of this there is a copy in the Rawlinson Collection at the Bodleian. There was a second reprint, also in 1730, with the heading :—

The Puerile Signs and Wonders of a Free-Mason (etc.) of which a copy exists in the Guildhall Library. The text with some small variations was published in the *Pennsylvania Gazette* of Dec. 5 to 8 in this year, a journal of which Benjamin Franklin was the Editor at the time.

* * *

Coming now to the Exposures in book-form we have :—

(Part II)

(1)

1724. The Grand Mystery of Free-Masons Discover'd . . . . . London; Printed for T. Payne near Stationers' Hall. 1724. Folio 12 pp. Reprinted at various times in last century, the most adequate reproduction being that issued by Mr. Carson, of Cincinnati, in 1867. There is a copy among the Rawlinson papers in the Bodleian.

(2)

1725. The Grand Mystery, the second edition. The imprint is: London; Printed for A. Moore, near St. Paul's. 1725. Folio, 20 pp. The copy in the Bodleian, which is also among the Rawlinson papers, and another in Dresden, were the only ones Dr. Chetwode Crawley knew of. To each edition there is a long title containing the assertion, which is stock form in these affairs, that the text is from papers found in the custody of a Free-mason who died suddenly. The second edition also contains " Two Letters to a Friend; The First Concerning the Society of Free-masons. The Second, Giving an Account of the Most Ancient Society of Gormogons," which are of some importance historically. The text was reproduced in Gould's *History*, in the Appendix to vol. III.

(3)

1730. Prichard's Masonry Disected. The title commences:
MASONRY / Disected; / being / A Universal and Genuine / DESCRIPTION / of / All its Branches, from the Ori- / ginal to this Present Time. /
The imprint is:

London : / Printed by *Thomas Nichols*, at the Crown, without *Temple Barr*. / mdccxxx.

This was also reprinted by Mr. Carson, of Cincinnati, in 1867. The work went through 21 editions by 1787, some of which, however, would seem to have been reprints in *Read's Weekly Journal* and the like. The earlier editions are extremely rare, only one or two of those up to the eighth being known apparently, with the exception of the fifth, of which no specimen at all is known to exist. The dates of the earlier editions are : I, as above, 1730; II, a reprint in Read, 1730; III, 1730; IV, 1731; V, n.d.; VI, 1736; VII, 1737; VIII, 1737. These two last have " a new and exact list of regular Lodges."

The work was also reprinted (the printer being J. Torbuck) with a different title and unimportant variations in 1737, the title now being :

The / Secrets of Masonry, / Made known to all Men, / By S. P. late Member of a / Constituted Lodge. / (etc.).

A work issued in Glasgow in 1803 by Robertson with the title : " The entertaining Mystery of freemasonry (etc.). By Sam. Pritchard " is also apparently a reprint of part of the original, and there is an earlier work with this same title which is no doubt of the same character, but the only copy known to me at present is in New Zealand. The 1803 work is also quite rare.

(4)

1738. Masonry farther Dissected.

This is a translation of a French exposure *L'Ordre des Franc Maçons trahi.*, but of an edition earlier than that of 1745 which has usually been considered the first. There is a very long title-page and an imprint : London : Printed for J. Wilford, ... Where may be had, Masonry Dissected. The Seventh Edition. Pr. 6d. 8vo, pp. 32.

Notwithstanding this announcement and the title chosen for the work, it has nothing in common with Prichard, with which indeed it is utterly at variance. Dr. Chetwode Crawley remarks (*Trans. of Lodge Quatuor Coronati*, ix, 84), " It is excessively rare, and fetches to-day, in open market, more than a hundred times its original price."

(5)
1754. The Freemasons examin'd. With a long title, and a preface explaining how the author came by the alleged secrets, his account of which is to replace Prichard, who is unworthy of credit. The whole thing is an elaborate skit. Of the original edition the Library of the Prov. Grand Lodge of Worcestershire possesses what appears to be the only copy known. Editions followed one another in rapid succession, but are all now very rare, the dates being: II, 1754, a copy in the B. M. and one other known in private ownership; III, ?, no copy known; IV, 1754, one copy in private hands; V, N.D. ? 1758; VI, N.D. ? 1758. (*Vide* article by Mr. J. T. Thorp in *Trans. of Q.C.*, XX, 96).

(6)
1759. The SECRETS of the Free Masons revealed by a disgusted brother. (etc.). Second edition; London: Scott. 8vo, 32 pp. The first edition does not appear to be known, and this is rare.

(7)
1760. Three distinct knocks. This also went through many editions during the century, several bearing no date, and was reprinted at Maidstone and in Ireland. The original publisher was Serjeant; the earlier editions 4to.

(8)
1762. Jachin and Boaz. London: Nicoll. 4to. Of this work there were twenty-six editions up to the Union of 1813. There were also American editions in 1793 and 1803, and even after the Union it continued to appear. The early editions are rare, the first few being very scarce indeed.

A reply was issued to *Jachin and Boaz* immediately on its appearance which is to-day even more of a rarity. It was entitled:—

A free-mason's ANSWER to the suspected author of a pamphlet entitled Jachin and Boaz, or An authentic key to freemasonry. Addressed to all masons, as well as to the public in general. London: Cooke, 1762.

A copy is in the British Museum, and a fifth edition is also known, published by Nicholl in 1764; but of the intermediate editions, or of later ones if there were any, no copies appear to have come to light.

(9)
1764. Hiram. A long title. Published in London by Griffin. 4to, plate, 96 pp. Reprinted in Belfast by Joy in 1765, and a second edition in London in 1766. All are extremely rare.

(10)
1765. Shibboleth. Dublin, Sleater, with a long title of the usual type. 4to, 52 pp. There was also an issue in London of the same date. Both are very rare.

(11)
1766. Mahhabone; or The Grand Lodge door open'd. London: Johnson and Davenport. With an exceptionally long title. There was a second edition in the same year.

(12)
1766. Solomon in all his Glory. This is another translation from the French, the original in this case being *Le Maçon demasqué*. The Library of Lodge Quatuor Coronati possesses a copy of this edition. A second edition, " with the addition of two beautiful copper-plates " was issued by Robinson and Roberts, London, in 1768. Wilkinson issued a reprint of this, with four copper-plates, in Dublin in 1777; 4to, viii, 72 pp.

(13)
[1777]. Tubal-Kain; being the second part of Solomon in all his Glory. With a very long-title, the only edition known being published by Wilkinson with the imprint: London, Nicoll. Reprinted Dublin; Wilkinson. There would therefore appear to have been an earlier edition. 4to, 32 pp. This and Mahhabone, No. 11 above, are the rarest of the whole series.

(14)
1824-5. The Cat out of the bag! Containing the whole secrets and mysteries of freemasonry never before devulged (*sic*). By Runt & Pitcher. London.
Published in four parts, but I do not know of any copies in libraries or museums, and Wolfstieg gives none. Some account of it will be found in the *Freemason*, vol. xlvii (1907-08).

(d) HISTORICAL

(1)

1746. THE Sufferings of JOHN COUSTOS for FREE-MASONRY (etc.). London: Printed by W. Strahan for the Author. (Also includes a long account of the Inquisition.) 8vo, 400 pp.
    Also issued in Dublin in the same year and a second edition with additional masonic matter was published at Birmingham in 1790. These are all of some rarity. Later editions at Hull in 1810, and by Spencer, London, in 1847.

(2)

[1764]. The complete Free Mason, or Multa Paucis for Lovers of Secrets. 4to, plate, 176 pp. Contains a history which although based on Anderson and the 1756 Book of Constitutions has important variations of its own. Also a list of Lodges which enables us to date the work; there is no date on the title-page. The late Mr. Whymper catalogued an edition of the previous year with 162 pages. Of extreme rarity.

(3)

1772. Preston's Illustrations of Masonry. A work that ran through seventeen editions up to 1861, being revised and brought up to date periodically by Oliver and others. It is only the original edition that is rare. 8vo, xxiv. 264 pp.

(4a)

1777. The Principles of Freemasonry delineated. Exeter. Printed (and sold) by R. Trewman, behind the Guildhall. MDCCLXXVII. With a plate of the medal of the Union Lodge at Exeter, This work consisted of a collection of charges and addresses appropriate to various occasions, or given at Exeter and elsewhere, together with an account of the proceedings at the dedication of Freemason's Hall and other matter. There was also a collection of songs, prologues and epilogues. It was published by subscription and there is a list of subscribers and a list of Lodges. In 1782 there appeared;

(4b)

1782. The Elements of Freemasonry delineated. Kingston, Jamaica: Printed by Brother William Moore (etc.). There is no plate, but the text, as far as the prose portion is concerned, is practically the text of the earlier work unaltered. Certain sections come in a different order, and some passages are left out; it also has its own list of subscribers and, what is of some interest, a list of the Lodges in Jamaica in 1781, with the names of their officers. The songs, etc. are however a different collection. Neither of these works is known to Wolfstieg. He gives however the third of the series, which is:

(4c)

1788. The Elements of Freemasonry delineated. By R. Ray. Liverpool: 1788. This is only known from a catalogue reference of 1861; no library appears to possess a copy. There was a second edition published at Belfast in 1808, which is also of considerable rarity; a copy is in the Library of Quatuor Coronati. The work is practically a reproduction of the publication of 1782, except for the omission of the lists of subscribers and local lodges; also there is now no locality mentioned for any of the prologues.

(5)

1797. The Freemason's Monitor, or Illustrations of Masonry. By a Royal Arch Mason. Albany: Spencer and Webb. 1797. 8vo.

The writer was Thomas Smith Webb, and the work is in two parts, but with continuous pagination, of which the first is merely a version of Preston's *Illustrations*, No. 3 above, the second being a description of the " ineffable degrees." There were many editions, and in the later ones additional matter was introduced relating to the history of the Craft in America. The original edition, pagination 1-216, 217-284, is of exceptional rarity. A reprint was issued in New York in 1899.

(6)

[1816]. Lectures on Masonry. W. Finch. No place or date of publication, but from what is known of Finch's career, the work can be dated with approximate accuracy.

(7)

[1816]. Lectures and Ceremonies of Freemasonry. W. Finch. 8vo.

As before, there is no date or place of publication, and the work is only dated by reference to the known facts of Finch's career. He is usually referred to as the masonic charlatan; he devised a system of Freemasonry of his own, for the imparting of which he exacted fees. The catalogue of 1861, referred to above (under No. 4), mentions both these publications.

(8)
1847. A pamphlet printed by William Platt, the Wor. Master of the Lodge of Friendship, No. 26, with a long title dedicating to the members of the Lodge " this cento of shreds and patches gleaned from the . . . eighteenth century," etc. A reprint of an address delivered in the Lodge. The contents are mainly biographical ; there is a copy in the Worcestershire Masonic Library.

(9)
1870. Masonic Lectures delivered in open Lodge, Chapter, etc., by R. W. Br. Col. Alexander Greenlaw (etc.). Madras, Higginbotham & Co., Publishers and Booksellers. 1870. Printed at the Asylum Press, Mount Road, by William Thomas. With a dedication to Earl Mayo, and preface. 8vo. ; pp. viii, 240.

Wolfstieg gives the publisher as Trübner, but they can only have published as London Agents for the Madras firm. He also gives the date with a query, but there is no doubt about it. He only knew of the work from a reference in a masonic periodical.

(10)
1871. Unpublished Records of the Craft. By William James Hughan, with valuable appendices (etc.). London: Kenning. 1871. 8vo. ; 54 pp.

Only fifty copies were printed. The great reputation Mr. Hughan subsequently attained as a masonic student has made all his early works to be eagerly sought after ; but this and No. A, ii, 17 have the added value of having been originally published in a very restricted edition.

(11)
1874. Memorials of the Masonic Union of A.D. 1813 (etc.). (A long title.) Compared and arranged by William James Hughan. London: Chatto & Windus. 1874. 4to ; Plate ; 119 pp.

The work also included (and the title referred to) Dassigny's *Serious and Impartial Enquiry*, No. F 6 *infra*. A reprint was issued with additional matter, by the Leicester Lodge of Research in 1913.

(12)
1884. Origin of the English Rite (etc.). By William James Hughan. A preface by T. B. Whytehead. London: Kenning. 1884. 8vo ; 4 plates ; vii, 150 pp.

In the preface to the second edition, of 1909, Mr. J. T.

Thorp of Leicester speaks of this first edition as follows:
" The limited edition was soon exhausted, and at the present time it is almost impossible to obtain a copy of the book, even at a very high price."

(13)

1893. Builders' Rites and Ceremonies. Two Lectures on the folk-lore of Masonry, delivered by G. W. Speth to the members of the Church Institute, Margate, on the 30th October and 13th November, 1893. Margate: Printed at Keble's Gazette Office. 1894. Paper covers; 8vo; 52 pp. Only 200 copies printed.

(14)

1907. A history of the Westminster and Keystone Lodge. J. W. S. Godding. Plymouth: Brenton & Son. Only 250 numbered copies printed.

There are numerous Lodge and local Histories, in many cases privately printed, and in limited editions. Mr. F. Leigh Gardner has catalogued all that he could trace in Vol. III of his *Catalogue Raisonné of works on the Occult Sciences* (1912), but it would serve no useful purpose to attempt, in this place, to discriminate between them in respect of their rarity, even if it could be done satisfactorily. Wolfstieg, however, specifies one work of the kind as rare, which I should perhaps give. It is:

(15)

1882. An attempt at compiling a History of Freemasonry in Stafford (etc.). By T. Ward Chalmers. Wright. Stafford.

Of this there is a copy in the Worcestershire Library.

(e) SERMONS AND SPEECHES

(1)

[1727]. A / SPEECH / Deliver'd to the / Worshipful and Ancient Society of / Free and Accepted Masons, / At a *Grand Lodge,* Held at *Merchant's- / Hall,* in the City of YORK, on St. John's / Day, *December* the 27th, 1726. / The RIGHT WORSHIPFUL Charles Bathurst, Esq., / Grand-Master. / By the *Junior Grand-Warden. / Olim Meminisse Juvabit.* / York: Printed for *Thomas Gent,* for the / Benifit of the Lodge. /

The only copy that has been traced so far is in the British Museum. The Speech attracted much attention at the time and there were several reprints in Cole's *Constitutions,* A (1) 4 *supra,* and elsewhere. It is also of considerable historical importance and has in more recent years been reproduced in Hughan's *Masonic Sketches* and in other works.

(2)

1750. Brotherly Love Recommended.

A sermon with a long title; preached at Boston, U.S.A., on 27th Dec. 1749, by Chas. Brockwell. Published at Boston in 1750, the printer being John Draper. The only copy known is in the British Museum. This was also reprinted in the Pocket Books.

(3)

1750. A Sermon preached at Gloucester on 27 Dec., and printed for the Author by Robert Raikes. The name of the preacher is unknown, and the only copy of the sermon itself is an imperfect one in the Library of Lodge Quatuor Coronati. *Vide* No. 231 of Mr. Dring's appendix already cited. 8vo; 24 pp. There was a second edition in 1752 of 30 pp.

(4)

1752. Masonry founded on Scripture, in a Sermon preached at Chatham on New Year's Day, 1752, by William Williams. London: 1752. 4to. Mentioned in the 1861 catalogue already referred to.

(5)

1757. The Light and Truth of Masonry explained. (With a long title.) By Thomas Dunckerley. London: Davey & Law. 8vo; 40 pp.

Two charges, one delivered at Plymouth on the occasion of the dedicating of a new Lodge room at the Pope's

Head Tavern, and the second at the same Lodge on the 24th June in that year. The Lodge Quatuor Coronati possesses what appears to be the only copy known. There was a second edition in 1758, 4to, 24 pp., which is also of great rarity.

(6)

1757. A Discourse upon Masonry. By George Minty. Dublin: Printed for the Author, by Alcx. M'Culloh, in *Skinner Row*, 1757.

With a very long title. The actual discourse was delivered in 1742 when the author was Master of a Lodge in England (which he does not specify). He appears to have embarked on the publication as a means of raising funds, and in 1772 he brought out a second edition, with the addition of "fraternal melody," the publisher being Wilkinson of Dublin, only on this occasion he stated that the oration had been delivered in that same year, the locality being unspecified. Both editions are quite rare.

At p. 104 of Vol. IX of the Transactions of Lodge Quatuor Coronati will be found comments on the work and its author by Mr. Conder, who also transcribes in full the title page to the first edition.

(7)

1766. The Excellency and Usefulness of Masonry (etc.). By Thomas Bagnall. London: Stuart. 8vo; pp. iv, 5-32.
There is a copy in a masonic Library at Hamburg.

(8a)

1768. FREEMASONRY / The High-way to Hell. / a / SERMON: / Wherein is clearly proved, / Both from Reason and Scripture ; That / all who profess these Mysteries are / in a State of Eternal Damnation. / (Two texts) / London : / Printed in the Year m,dcc,lxviii. /
8vo ; 22 pp.

(8b)

1768. The same title, but "eternal" omitted, and the imprint is: London: Printed for Robinson and Roberts, at No. 25, in Paternoster Row, m,dcc,lxviii.
8vo ; 39 pp.

Except in respect of pagination and title, these are identical works and it is not possible to say which appeared first. Robinson and Roberts were the publishers of C 12 *supra*. The so-called Sermon was in practice a scurrilous pamphlet, and is quite unlikely to have been ever delivered. The writer is unknown. A second edition was published on May 2nd, in this same year, also by Robinson and Roberts, "and sold by R. Goadby in Sherborne." There was a reprint by W. G. Jones and J. Millikin, at Dublin, also in this same year. There was further a German translation now and a French in 1769. The pamphlet has been reprinted with an introduction as No. V of the Leicester *Masonic Reprints* (1922). No Masonic Library appears to possess a copy of any edition. Wolfstieg gives the date of the first edition as 1761 (his No. 3598), but this is an error.

(9*a*)

1768. MASONRY / the / Turnpike-Road / to / Happiness in this Life / and / Eternal Happiness hereafter. / Dublin : / Printed by James Hoey, senior, at the/ *Mercury*, in Skinner-Row. m dcc lxviii. /
8vo ; 32 pp.

(9*b*)

1768. The same, but the imprint is : London : Published April 18, 1768. Printed for S. Bladon in Paternoster Row and sold by R. Goadby in Sherborne.

These again are duplicates. The former is reprinted with No. 7 *supra* in No. V of the Leicester Reprints. The pamphlet is a reply to the Sermon by an unknown author. There is a copy of the Dublin edition at Leicester. There was a German translation published at Frankfort in 1769. The Sermon provoked three other rejoinders, all of great rarity to-day. They are :—

(10)

1768. Remarks on a Sermon lately Published (etc.). By John Thompson. London : Printed by S. Axtell and H. Hardy, for T. Evans, at No. 20, in *Pater-noster Row*. mdcclxviii. (Price One Shilling.) 8vo ; 35 pp.

(11)

1768. Masonry Vindicated. A sermon (with a very long title). London : printed for J. Hinton. 1768. 8vo ; 35 pp.

(12)

1768. An Answer to a certain Pamphlet lately published under the solemn Title of " A Sermon, or Masonry the Way to Hell." By John Jackson. Philanthropos. 1768.

In the Leicester Reprint Mr. J. T. Thorp, in the introduction, observes that of the whole series there are probably not more than a dozen copies in existence. These last three are unknown to Wolfstieg.

(13)
1776. An Oration. Delivered at the Dedication of Freemason's Hall on Thursday May 23 1776. By William Dodd. Published by general request under the sanction of the Grand Lodge. London: Robinson. 1776.

The Hall referred to is the present building in Great Queen Street. The speech was frequently reprinted in miscellanies, but copies of the original publication are scarce. There is one at Worcester.

(14)
1808. Orations of Fred. Dalcho. Reprinted by permission of the author under the sanction of the Ill. the College of Knights of K.H. and the Original Chapter of Prince Masons of Ireland. Dublin: King.

Of extreme rarity; a copy in the Library of Lodge Quatuor Coronati, and another at Worcester.

(f) MISCELLANEOUS

(1)

[? 1722] THE Free-masons, an Hudibrastic Poem. 8vo, 24 pp. with a long title. The second edition of this with date 1723 is in the B. M.; the advertisement of this second edition appeared in the *Daily Post* of Feb. 15 of that year, as well as in other contemporary journals. (Robbins in *Trans. Q.C.* xxii, 75). Wolfstieg dates the first edition 1722, but Begemann states (*History*, ii, 173) that the work first appeared in 1723. These two editions were " Printed for A. Moore, near St. Paul's "; a third edition was published in 1724 by Warner.

(2)

1725. The / *Freemasons Vindication,* / being an / ANSWER / To a Scandalous Libel, entituled the *Grand Mistery* / of the *Free Masons* discover'd &c. / wherein is plainly prov'd the falsity of that / Discovery, and how great an imposition it is on the Publick. / . . .

A foolscap broadside; there is a copy in the British Museum, and another in the Rawlinson papers at the Bodleian. It is an answer to No. C ii, 1.

(3)

1726. The Freemasons Accusation and Defence. 8vo, 39 pp. with a long title. Printed for J. Peele. . . . . and N. Bradford. Advertised in Jan., and a second edition in March of that year. Wolfstieg also gives a third edition but without any details. There is a copy in the B. M.

(4).

1726. A Full Vindication etc. By a Lover of Harmony and Good Fellowship. London: Printed for J. Roberts in Warwick Lane, 1726. 8vo, 27 pp. A reply to the previous work. The only copy known is in the Bodleian.

(5)

1726. An Ode to the Grand Khaibar . . . . London: Printed and Sold by J. Roberts in the Oxford Arms Passage near Warwick Lane. 4to, 9 pp. The only copy known is in the Library of the Quatuor Coronati Lodge. It is written to bring into contempt the history and poems in the Book of Constitutions of 1723.

(6)

1730. A New Model For the Rebuilding Masonry etc. By Peter Farmer. Dedicated to Mr. Orator Henley. Printed for

J. Wilford. 32 pp., of which the last 16 are songs. The only copy known is in the B. M.

(7)

1731. A / DEFENCE / of / MASONRY, / Occasioned by a Pamphlet, / called / Masonry Dissected. / / *Rarus Sermo illis, & magna Libido* Tacendi.
Juv. Sat. 2. / /
London : / Printed for J. Roberts, near the *Oxford-Arms,* / in *Warwick-Lane.* / M.DCC.XXXI.
Until within a few years ago this work was only known from the advertisements of its publication in the *Daily Post* and *Daily Journal* of Dec. 15th and 16th respectively, 1730. There is now a copy in the Grand Lodge Library. Collation : 8vo ; half-title ; title ; pp. 1-27. It was reprinted in the *Constitutions* of 1738 with the same title, as also in Smith's Freemason's Pocket Companion of the same year. (B. 4.) But it is noteworthy that in the *Constitutions*, it is described as published in 1730, as in fact it was, whereas the date on the title-page is 1731.

(8)

1736. The Book M : or Masonry Triumphant. In two parts with a long title. Newcastle-upon-Tyne.: Printed by Leonard Umfreville and Company. The first part is a history, etc. ; the second consists of songs and poems with a list of meeting places of Lodges. It is in fact on the lines of the Pocket Companions. 8vo, x, 76 ; 66, x pp. There is a copy in the Masonic Library at Leeds.

(9)

1744. A Serious and Impartial ENQUIRY Into the Cause of the present Decay of FREEMASONRY in the Kingdom of Ireland . . . . By Fifield Dassigny M.D. . . . Dublin : Printed by Edward Bate in George's-lane near Dame-Street. 4to, 80 pp. Only three copies known to exist ; one is in the Library of the G. L. of Iowa, and another at Leeds.

(10)

1758. A Collection of Freemasons' songs with complete list of all the regular Lodges both in England and Scotland down to the year 1758 by James Callendar. Edinburgh : Donaldson.

(11)

1765. A Defence of Freemasonry, (etc.). A long title, the imprint being : London : / Printed for the Author, and sold by W. Flexney, near / *Gray's-Inn Gate, Holborn ;* and

E. Hood, near *Stationers- / Hall, Ludgate-Street.* 1765. / (Price One Shilling.) 4to, 64 pp. This is a reply to Dermott's attack on the Grand Lodge and the history as put forward by Anderson and his followers. There is a copy in the Library of Grand Lodge ; there *may* be others in private ownership. Reproduced in facsimile in Sadler's *Masonic Reprints,* 1898.

(12)

1773. Fraternal Melody. By Will. Riley. London : printed for the Author, in Great James Street, Bedford Row, Holborn, mdcclxxiii. (Price Two Shillings.) With a very long title; gives songs, etc., for the use of a number of different friendly societies. There is a copy in the Library of Grand Lodge.

(13)

1775. An Introduction to Freemasonry. In four parts ; W. Meeson ; Birmingham. 8vo, 100 pp. Published by Pearson and Rollason. Reprinted in London by Baldwin in 1776. A long title ; a copy exists in the Library of the Worcestershire Provincial G.L.

(14)

1783. The Use and Abuse of Freemasonry ; a work of the greatest utility to the brethren of the Society, to mankind in general, and to the ladies in particular. By George Smith. London : Kearsley. 4to ; xxvii, 399 pp.

There is a copy at Hamburg. It contains, among other matters, a reprint of Dodd's Oration, No. E 13.

(15)

1788. The Institutes of Freemasonry ; to which are added a choice collection of epilogues, songs, etc. Addressed to the Sea Captains' Lodge. Liverpool : Johnson. 8vo, x, 266 pp. A work on the lines of the Pocket Companions but with additional matter.

(16)

1790. The Philosophy of masons in several epistles from Egypt to a nobleman. London : Ridgway. 8vo, x, 265 pp. By Thomas Marryat. This provoked a reply by H. E. Holder, published by Pine, Bristol, in 1791 (8vo, 22 pp.) ; and to this in its turn an anonymous layman rejoined in a " Letter to H. E. Holder (Bristol : Routh, 1791 ; 8vo, 11 pp.) ; to which Holder retorted with "An Answer to the layman's letter " (Bristol : Pine 1791. 8vo, 8 pp.). The whole set is rare.

(17)

[? 1794]. The Freemason's Repository, with songs, odes, etc., and the secret way of writing used among Masons. Birmingham: Printed by and for J. Sketchley, Auctioneer, No. 139, Moor Street, n.d. 8vo. As Sketchley got into financial difficulties in 1794 the book is not later, and Kloss considered it was published in 1786. There was a second edition in 1812. (cf. Note by Mr. J. T. Thorpe in *Trans. Q. C.* xviii, 147).

(18)

1799. The masonic Museum. Containing a select collection of the most celebrated songs, sung in all respectable Lodges, with a complete list of the Lodges of Instruction. London: Roach.

There was a second edition in 1801; there is a copy of the first at Worcester.

(19)

1812. An Enquiry into the late disputes among the Freemasons of Ireland (etc.). Belfast: Printed by Joseph Smyth, 115 High Street. 1812.

With an extraordinarily verbose title, which will be found in full at p. 58 of Vol. X of the Transactions of Lodge Quatuor Coronati. Dr. Chetwode Crawley only knew of two copies, when describing the work in 1897.

(20)

1818. Masonic Melodies; being a choice selection of the most approved masonic songs etc., etc., the whole set to music (etc.). By Luke Eastmann. Boston: 1818.

There is a second edition of 1825, but it is not of the same rarity.

(21)

1880. The medals of the Masonic Fraternity described and illustrated. By William T. R. Marvin. Boston: 1880. 4to; x, 329 pp.; 18 plates. Only 160 copies were issued.

(22)

1891. A Catalogue of Bibliographies, lists and catalogues of works on Freemasonry. Compiled by H. J. Whymper. London: 4to; 16 pp. Only 100 copies issued.

* * *

In conclusion a word may perhaps be said as to spurious

books. The non-existent First Part of the *Constitutions* of 1815 has already been referred to; but one occasionally sees mention made of an edition of the *Constitutions* printed at Brussels in 1722. There is no such work. Two works of dates prior to 1722 are given by Kloss; they are *A Short Analysis of the unchanged rites and ceremonies of Freemasons* said to be printed for Stephen Dilly in 1676, and *Observations and Enquiries relating to the brotherhood of the Freemasons*, supposed to be written by Simeon Townsend and published in 1712. They have not been traced and the dates assigned to them make it unlikely that they ever will be. A list of all such bibliographical references is given by Mr. Dring at the end of the Appendix to his Inaugural Address, to which reference has already been made, and to which I would once more express my indebtedness as to many items in the present compilation. I would also wish to acknowledge the assistance and information given me by Mr. W. J. Songhurst, the Secretary of the Quatuor Coronati Lodge, and the help rendered by putting at my disposal many works in the Lodge Library.

MAGGS BROTHERS

34 & 35 CONDUIT STREET
NEW BOND STREET
LONDON, ENGLAND

Carry one of the largest and choicest stocks
in England of

Fine and Rare
BOOKS, PRINTS
and
AUTOGRAPHS

Illustrated Catalogues in each Department
regularly issued

These Catalogues appeal especially
to the Connoisseur
Collector and Antiquarian

*Customers "Desiderata" searched for and reported
free of charge*

Items of Rarity and Interest always gladly
Purchased

SHIPMENTS to AMERICA EVERY WEEK

JUST PUBLISHED

THE CONSTITUTIONS OF THE FREEMASONS 1723

An absolute facsimile reproduction of the FIRST EDITION of ANDERSON'S CONSTITUTIONS with an Introduction by Lionel Vibert, I.C.S. (retd.), P.M. of the Lodge of the Quatuor Coronati, London.

Post 4to., pp. XLVI, 92, boards, linen back, gilt top
£1 1s. net.

This is the first time that this rare and important work has been reproduced in exact facsimile. Only a limited edition will be printed.

BERNARD QUARITCH, LTD
11 Grafton Street, London, W.1

Messrs. P. J. & A. E. DOBELL
beg to announce
that they have always in stock a large collection of books on Freemasonry especially the early and rare books, such as the Constitutions, Eighteenth Century Tracts, etc., etc.

Also rare books in all branches of literature.

Catalogues free on application.

P. J. & A. E. DOBELL
8 Bruton Street, New Bond Street
London, W.1

Rider's Masonic Publications

A NEW ENCYCLOPAEDIA OF FREEMASONRY
(Ars Magna Latomorum)
And of Cognate Instituted Mysteries, Their Rites, Literature, and History

By *Arthur Edward Waite, P.M., P.Z., Past Senior Grand Warden of Iowa, etc.* Author of *"The Secret Tradition in Freemasonry," etc., etc.* With 16 full-page plates and many Illustrations in the text. 2 volumes. Royal 8vo (pp. xxxii+458, and pp. iv+488. Blue Cloth Gilt. 42s. net.

"There is evidence to spare in Mr. Waite's erudite pages of the folly and obstinacy of those people who have sought to make their own paths to the Temple Gate without the key which alone can open a passage through the mystic labyrinth which guards the upper slopes of the sacred mountain. There are others who really seek. . . For such as these this 'New Encyclopaedia' will be useful."—*The Times.*

THE MEANING OF MASONRY
By *W. L. Wilmshurst, P.M.* Demy 8vo, cloth, 10s. 6d. net.

CONTENTS.—Introduction—The Position and Possibilities of the Masonic Order—The Deeper Symbolism of Masonry—Masonry as a Philosophy—The Holy Royal Arch—The Relation of Masonry to the Ancient Mysteries.

"Among Freemasons it will be conceded that the author has justified the title of his work. The book is distinguished by patient research, erudition, lofty ideals, and good literary style. . . Mr. Wilmshurst carries one far beyond the ordinary conception of the Masonic Art."—*Yorkshire Post.*

MASONIC LEGENDS AND TRADITIONS
By *Dudley Wright.* Crown 8vo, cloth. Illustrated Frontispiece. 5s. net.

"A book acceptable alike to the trifler and the student."—*The Co-Mason.*

WOMEN AND FREEMASONRY
By *Dudley Wright.* Crown 8vo, cloth. 6s. net.

"To students of rituals certain of Mr. Wright's pages will prove of great interest."—*The Times.*

ROMAN CATHOLICISM AND FREEMASONRY
By *Dudley Wright.* Crown 8vo, cloth. 10s. 6d. net.

This is a historical not a controversial work. The facts are given without embellishment; they speak for themselves. The range covered by the work extends over two centuries, beginning with the latter part of the seventeenth century and carried up to the present day.

IN THE PRESS

FREEMASONRY: ITS AIMS AND IDEALS
By *J. S. M. Ward, B.A., F.S.S., F.R.Econ.S.* Author of *"Freemasonry and the Ancient Gods,"* etc. Demy 8vo, cloth. 10s. 6d. net.

CONTENTS: Part I.—The Ideals of Freemasonry—The Political Ideal—The Social Ideal—The Ritualistic Ideal—The Archæological Ideal—The Mystical and Religious Ideal—What Mysticism is—Conclusions. Part II.—Grave Problems:—Women and Masonry—The "Black" Lodges—The International Aspect of Freemasonry—Freemasonry and Established Religions—The Grand Ideal—The Formation of an International Grand Lodge.

WILLIAM RIDER & SON, LTD
8 Paternoster Row, London, E.C.4

Spcl
HS
395
V534
1923

BROCK UNIVERSITY
ST. CATHARINES, ONTARIO

LIBRARY

FOR USE IN SPECIAL COLLECTIONS ONLY

Milton Keynes UK
Ingram Content Group UK Ltd.
UKHW022014120124
435957UK00005B/116